YOUR KNOWLEDGE HAS VALUE

- We will publish your bachelor's and master's thesis, essays and papers

- Your own eBook and book - sold worldwide in all relevant shops

- Earn money with each sale

Upload your text at www.GRIN.com and publish for free

Nazim Nariman

Finite element analysis of the buckling critical loads in un-braced steel frames with multiple slenderness ratio configurations

GRIN Verlag

Bibliografische Information der Deutschen Nationalbibliothek:

Die Deutsche Bibliothek verzeichnet diese Publikation in der Deutschen National-
bibliografie; detaillierte bibliografische Daten sind im Internet über http://dnb.d-
nb.de/ abrufbar.

Imprint:

Copyright © 2013 GRIN Verlag GmbH
Druck und Bindung: Books on Demand GmbH, Norderstedt Germany
ISBN: 978-3-656-51293-6

Title

Finite element analysis of the buckling critical loads in un-braced steel frames with multiple slenderness ratio configurations

Author

Nazim Abdul Nariman

Nazim Abdul Nariman , PhD candidate - Institute of Structural Mechanics, Bauhaus Universitat Weimar – Germany.

Abstract

In this paper, two types of steel frames, steel frame without side sway permission and another with side sway permission are created in Abaqus with 10 multiple slenderness ratio of the columns by changing the length every time starting from 1 M and ending with 10 M length of the columns, Twenty models of steel frames with single story and single bay were created, the models are with the same 2D dimensions and material properties, the cross section of the steel is (0.5*0.5) M ,and the supports are fixed, two equal forces P= 1000 N are exerted on the frames in the position mentioned in fig 6, a beam section was defined for the frame integrated before analysis with Young modulus of elasticity $E=1*10^7$ N/M^2 · and shear modulus $G = 3.8*106$ N/M2 and poisons ratio $v = 0.3$.   .

A linear perturbation step is created for buckling and 10 eigenvalues are requested for analysis, a standard quadratic beam element type is generated with global seeding of 0.6, and 20 Jobs are created for every situation and conclusions have been obtained, the critical buckling loads of the frames fall in the ranges between the Euler loads forms which has been proved for each type of frames and this scientific approach was verified in this research, in addition to that the relation between the length of the column and the eigenvalues that represent the critical loads of buckling verified, and the simulations of the mode shapes of buckling of the steel frames were identified adopting finite element analysis which shows the amount of loads necessary to reach each mode shape of buckling for each type of steel frames mentioned before .

Keywords

Euler column, stiffness matrix, critical buckling load, eigenvalues and eigenvectors

Introduction to buckling

If a beam element is under a compressive load and its length if the orders of magnitude are larger than either of its other dimensions such a beam is called a column. Due to its size its axial displacement is going to be very small compared to its lateral deflection called buckling.

Slender or thin-walled components under compressive stress are susceptible to buckling and is called "Euler buckling" where a long slender member subject to a compressive force moves lateral to the direction of that force, as illustrated in Figure 1. The force, F, necessary to cause such a buckling motion will vary by a factor of four depending only on how the two ends are restrained. Therefore, buckling studies are much more sensitive to the component restraints that in a normal stress analysis. The theoretical Euler solution will lead to infinite forces in very short columns, and that clearly exceeds the allowed material stress. Thus in practice, Euler column buckling can only be applied in certain regions and empirical transition equations are required for intermediate length columns. For very long columns the loss of stiffness occurs at stresses far below the material failure.

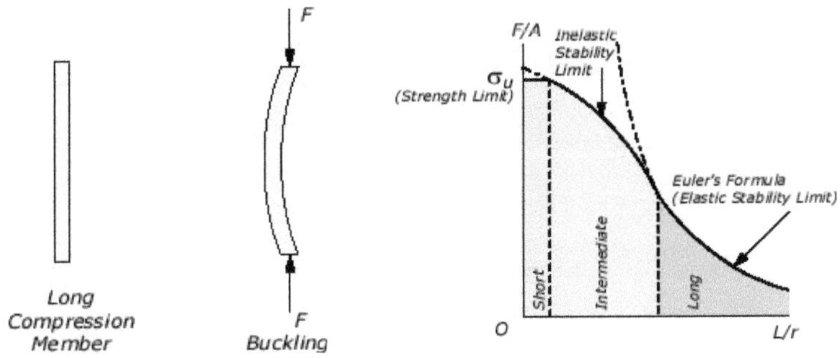

Figure.1 Long columns fail due to instability

Quite often the buckling of column can lead to sudden and dramatic failure. And as a result, special attention must be given to design of column so that they can safely support the loads.

Buckling can be related to the singularity of the tangent stiffness matrix, which in turn consists of two parts. The first part is the material stiffness matrix which is related to the deformational stiffness of the components, taking into account the connectivity of components in the current geometric configuration of the structure. For linear elastic components, the material stiffness is identical to the linear elastic stiffness, but updating the structural geometry to include the effect of any displacements. The second part is the geometric stiffness matrix, which is related to the component forces, and in some cases to the applied loading, taking into account the effect of a change in geometry from the current configuration. For typical structures, the material stiffness is positive for all deformation modes, mathematically referred to as positive-definite, whereas the geometric stiffness can admit negative values for certain modes, depending on the component forces and applied loading. It is therefore the effect of a negative geometric stiffness that can lead to a singular overall tangent stiffness matrix, and hence buckling.

Stability concept

The question of the stability of various forms of equilibrium of a compressed bar can be investigated by using the same theory as used in investigating the stability of equilibrium configurations of rigid-body systems (Timoshenko and Gere, 1963). Consider three cases of equilibrium of the ball shown in Figure.2. It can be concluded that the ball on the concave spherical surface (a) is in a state of stable equilibrium, while the ball on the horizontal plane (b) is in indifferent or neutral equilibrium. The ball on the convex spherical surface (c) is said to be in unstable equilibrium.

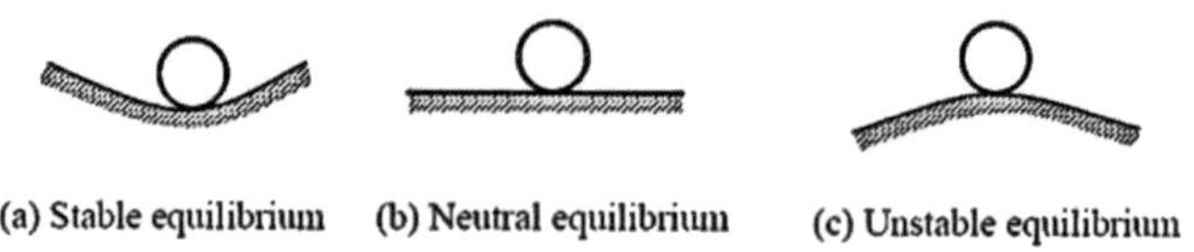

Figure.2 Ball equilibrium cases

The compressed bar shown in Figure.3 can be similarly considered. In the state of stable equilibrium, if the column is given any small placement by some external influence, which is then removed, it will return back to the un-deflected shape. Here, the value of the applied load P is smaller than the value of the critical load Pcr. By definition, the state of neutral equilibrium is the one at which the limit of elastic stability is reached. In this state, if the column is given any small displacement by some external influence, which is then removed, it will maintain that deflected shape. Otherwise, the column is in the state of unstable equilibrium.

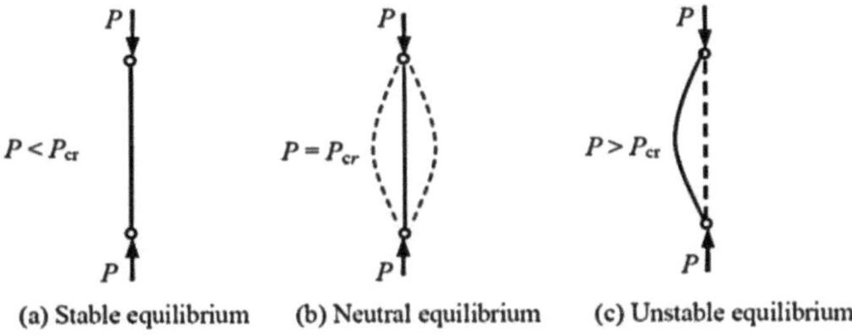

(a) Stable equilibrium   (b) Neutral equilibrium   (c) Unstable equilibrium

Figure.3 Column equilibrium cases

Euler Column

The Euler column is the axially loaded member shown in Figure.4 which is very idealized and is assumed to have a constant cross sectional area and to be made of homogeneous material. In addition, four assumptions are made:

1- The ends of the member are simply supported. The lower end is attached to an immovable hinge, and the upper end is supported so that it can rotate freely and move vertically but not horizontally.
2- The member is perfectly straight, and the load is applied along its Centroid axis.
3- The material obeys Hooke's law.
4- The deformations of the member are small enough so that the curvature term can be approximated to derive the Euler load P which is the buckling load.

Due to imperfections no column is really straight. At some critical compressive load it will buckle. To determine the maximum compressive load (Buckling Load) we assume that buckling has occurred as shown in Figure.4.

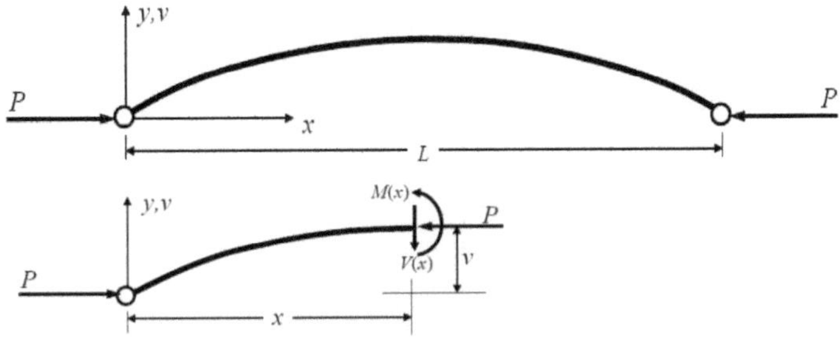

Figure.4 Buckling load

Equating moments at the cut end:

$$\Sigma M = 0 = Pv + M(x) = 0 \qquad \therefore M(x) = -Pv$$

But since the deflection of a beam is related with its bending moment distribution, then:

$$EI\frac{d^2v}{dx^2} = -Pv$$

which simplifies to
$$\frac{d^2v}{dx^2} + \left(\frac{P}{EI}\right)v = 0$$

where P/EI is a constant. This expression is in the form of a second order differential equation of the following type:

$$\frac{d^2v}{dx^2} + \alpha^2 v = 0$$

Where:
$$\alpha^2 = \frac{P}{EI}$$

The solution of this equation is:

$$v = A\ cos(\alpha x) + B\ sin(\alpha x)$$

Where A and B are constants, which can be determined using the column's kinematic boundary conditions.
Kinematic Boundary Conditions

$$\textbf{at } x = 0,\ v = 0:\ 0 = A + 0,\ \textbf{giving that } A = 0$$

$$\textbf{at } x = L,\ v = 0,\ \textbf{then: } 0 = B\ sin(\alpha L)$$

If $B = 0$, No bending moment exists, so the only logical solution is for: $sin(\alpha L)=0$ and the only way that this can happen is if :

$$\alpha L = n\pi, \qquad \text{where } n = \text{integer}$$

But since:

$$\alpha^2 = \frac{P}{EI} = \left(\frac{n\pi}{L}\right)^2$$

Then we get that buckling load as:

$$P = n^2 \frac{\pi^2 EI}{L^2} \qquad \dots\dots\dots\dots (1)$$

The values of 'n' define the buckling mode shapes, as in Figure.5:

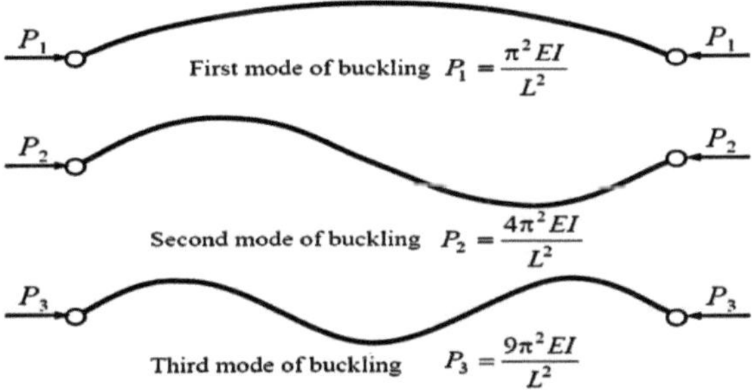

Figure.5 Buckling modes

However, since $P1 < P2 < P3$, the column buckles at $P1$ and never gets to $P2$ or $P3$ unless bracing is placed at the points where $v = 0$ to prevent buckling at lower loads. The critical load for a pin ended column is therefore:

$$P_{Crit} = \frac{\pi^2 EI}{L^2} = P_E \qquad \dots\dots\dots\dots (2)$$

Which is also called *Euler Buckling Load,*

PCrit   Critical or maximum axial load on the column just before it
begins to buckle

E Young's modulus of elasticity

I *least* second moment of area for the column's cross sectional area

L unsupported length of the column, whose ends are pinned

Critical Buckling Load

Buckling is that mode of failure when the structure experiences sudden failure when subjected to compressive stress. When a slender structure is loaded in compression, for small loads it deforms with hardly any noticeable change in the geometry and load carrying capacity.

At the point of critical load value, the structure suddenly experiences a large deformation and may lose its ability to carry load. This stage is the buckling stage.

The critical buckling load for a pin-pinned column (which is called the Euler buckling load) is given by the formula:

$$P = \frac{n^2 \pi^2 EI}{l^2}$$

n = 1,2,3 integer which is representing the mode of buckling for the simply supported columns.

Where n=1 for the lowest critical buckling load so the formula would be:

$$P_{cr} = \frac{\pi^2 EI}{l^2}$$

So the critical loads at which nonzero deflections are possible are the eigenvalues, and the deflected shapes that can exist at these loads are the eigenvectors. The smallest eigenvalue is the critical load and the corresponding eigenvector is the buckling mode shape.

Buckling of frames

By considering a single story frame with a single bay, and for the buckling study we assume the external loads P to act directly over the columns so that there is no bending moment in any member of the frame prior to buckling, and the frame is categorized to two types: frames where side sway prevented and frames with side sway permitted.

We consider the first type of frames in which side sway is prevented. At the critical load the frame buckles as it mentioned in the figure. The buckling takes place when the applied load P is equal to the critical load of the columns, the upper end of each column is elastically restrained by the beam to which the column is rigidly connected, and that the critical load of the column therefore depends not only on the column stiffness, but also on the stiffness of the beam. We when we assume that the beam is infinitely flexible, the beam then is unable to offer any rotational restraint to the upper ends of the columns as shown in Figure.6.

In this case the columns behave as if they were fixed at one end and hinged at the other, and the critical load of the frame is approximately equal to twice the Euler load of the columns.

For an actual frame the critical load in which side sway is prevented, can be bracketed as follows:

$$2P_\varepsilon < P_{cr} < 4P_\varepsilon \qquad \ldots\ldots\ldots\ldots\ldots (3)$$

When we consider the second types of frames in which side sway is permitted, and the base is fixed and it is assumed that the frame's material behaves according to Hooke's law, that the deformations remain small, and that there is no primary bending present in the frame prior to buckling, and the shear forces that arise from the bending of the horizontal member are neglected, and if we assume the beam to be infinitely flexible, the upper ends of the columns are free to both rotate and translate as shown in Figure.11.

In this case the columns act like to be fixed at the base and free at the top, and the critical load of the frame is equal to one fourth the Euler Load of the columns.

The critical load of the frame whose upper joints are free to translate laterally must therefore lie between P_ε and $1/4\ P_\varepsilon$. That is:

$$1/4\ P_\varepsilon < P_{cr} < P_\varepsilon \qquad \ldots\ldots\ldots\ldots\ldots (4)$$

Buckling strength

Designs based on the alignment chart are reasonably accurate only when all the individual columns in a story buckle simultaneously under their individual proportionate share of the total gravity load. The columns cannot brace each other in this situation — their total strength is required to support their own gravity loads, leaving no reserve which might be counted upon to provide a bracing force for other columns.

There are situations in which the individual columns have excessive buckling strengths. If the two exterior columns contain axial loads such that the buckling load of these columns is not reached when the interior columns reach their independent buckling loads, the system will not buckle. This may occur when different loading conditions govern the design of various columns in a story. Shear resistance will be developed in the exterior columns which counteracts the side sway tendency. (If all columns want to buckle simultaneously, there will be no shear resistance available. the stabilizing effect of the lightly loaded exterior columns buckling will occur. The critical load for the interior columns is increased and their effective length is decreased. The stabilizing effect can be such that the effective length of some of the columns could be reduced to 1.0, even though there is no apparent bracing system.

It is safe to treat separately each column to which beams are rigidly attached and to use the alignment chart to get the individual strengths. However, in some instances this usual approach may be unduly conservative. Side sway buckling is a total story phenomenon. A single individual column cannot fail by side sway without all the columns in the same story also buckling in a sway mode. On the other hand, buckling in a non-sway mode is an individual phenomenon. Each column's non-sway buckling load is reasonably independent of the buckling load of the other columns.

Buckling modes of frames

1- Symmetric buckling

If the frame is prevented from translating laterally at the top, buckling will occur in the symmetric mode, as indicated in Figure.6 .The critical load for a fixed base portal frame whose beam has the same stiffness as the columns and that is laterally restrained is:

$$P_{cr} = 25.2 \frac{EI}{L^2} \qquad \cdots\cdots\cdots\cdots (5)$$

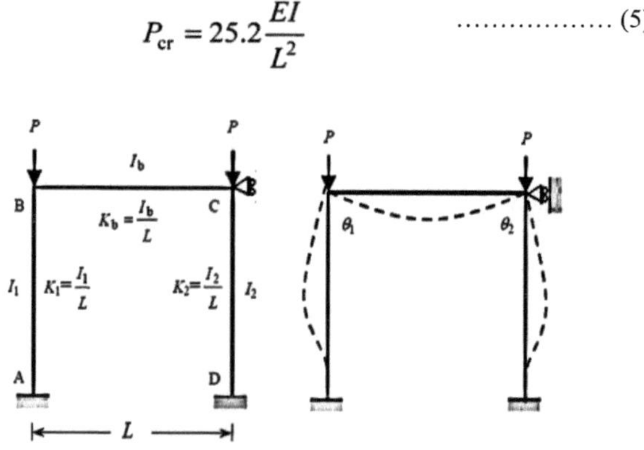

Figure.6 Symmetric buckling

The mode shapes of buckling of the frame when side sway is prevented can be seen very apparent in the output of Abaqus jobs indicating the magnitude of critical load of buckling (eigenvalues) associated with each mode shape in the Figure.7-a till Figure.7-k.

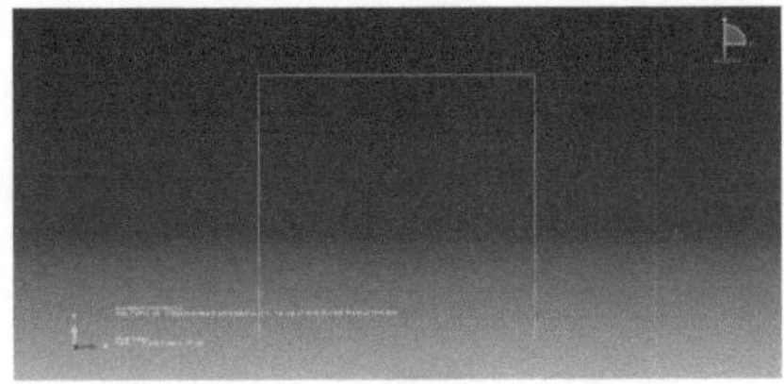

Figure.7- a Un-deformed shape

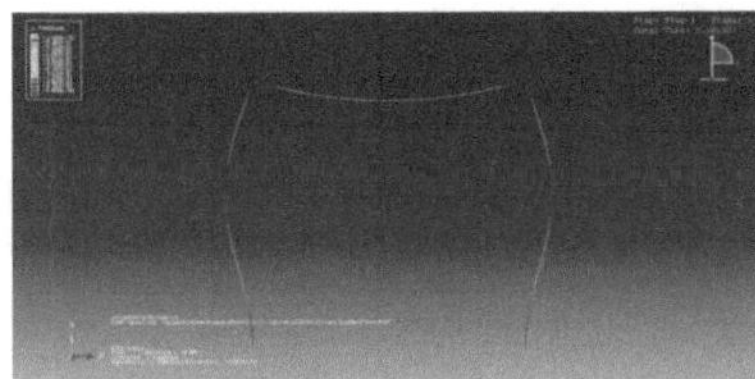

Figure.7- c Mode shape 2

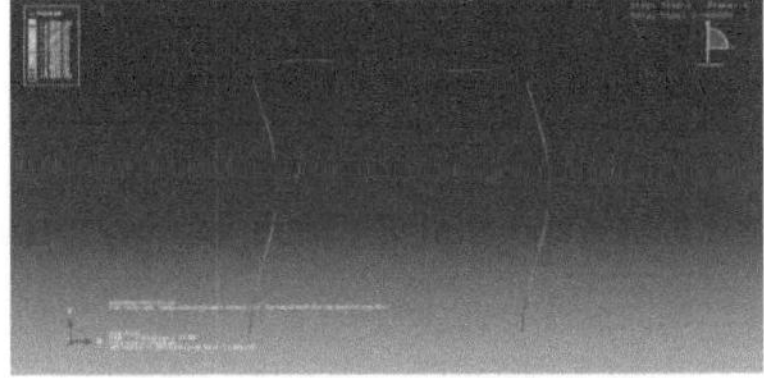

Figure.7- b Mode shape 1

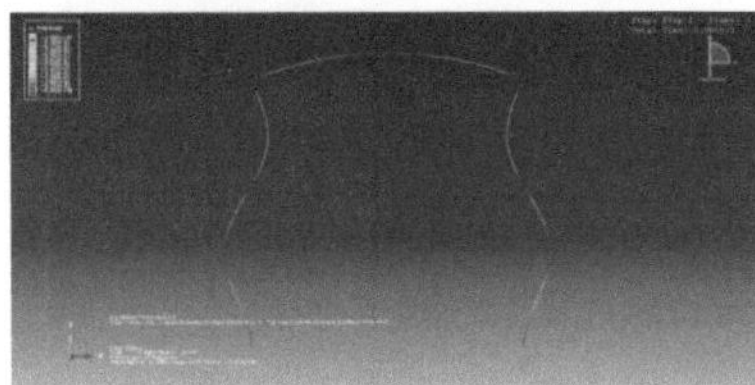

Figure.7- d Mode shape 3

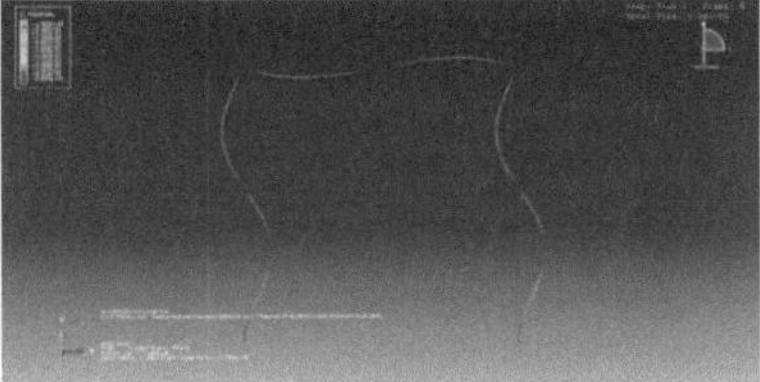

Figure.7- e Mode shape 4

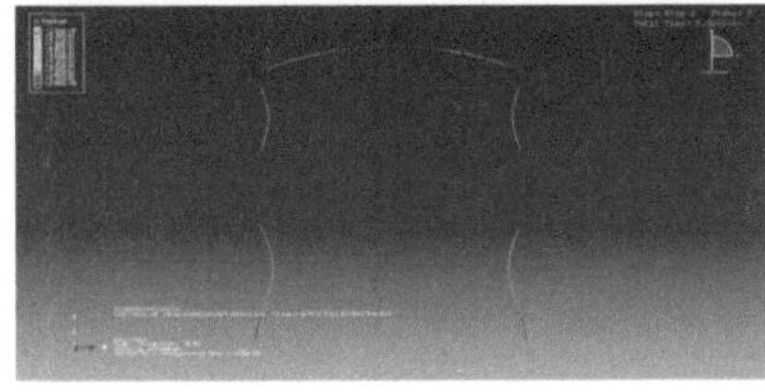

Figure.7- f Mode shape 5

Figure.7- g Mode shape 6

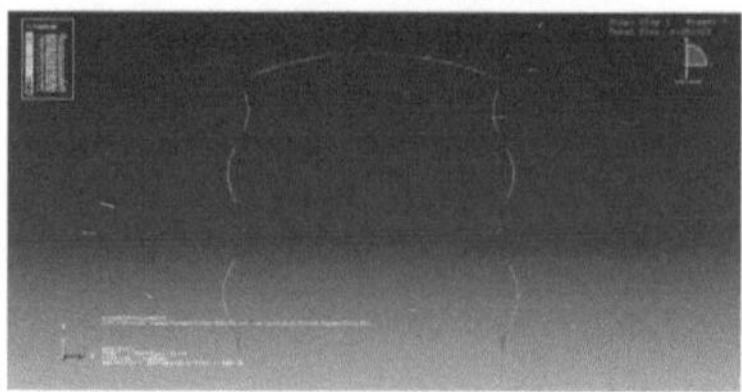

Figure.7- h Mode shape 7

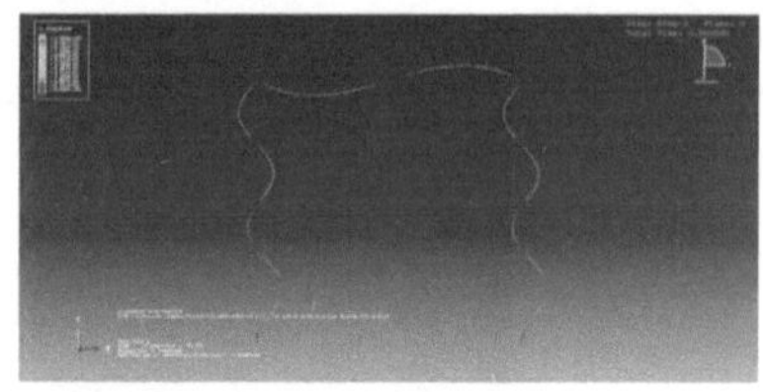

Figure.7- i Mode shape 8

Figue.7- j Mode shape 9

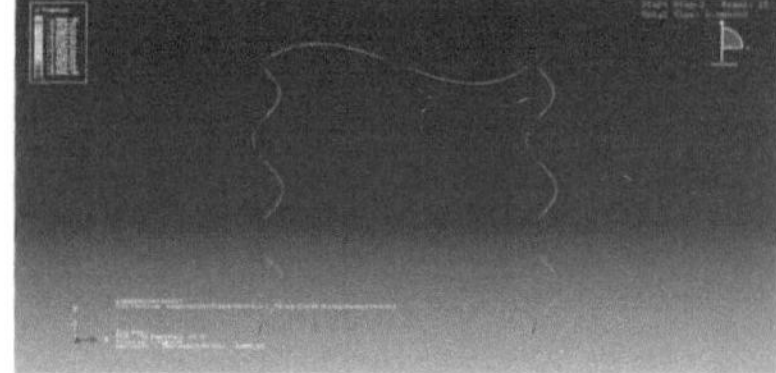

Figure.7- k Mode shape 10

Table -1 is the results of Euler loads and Critical buckling loads for 10 situations of column length for the frame with side sway prevented.

Table – 1 Euler Loads and Critical Loads

π2	E N/M2	I M4	L M	PE N	Pcr N
9.877	10000000	0.0052	1	513604	1310400
9.877	10000000	0.0052	2	128401	327600
9.877	10000000	0.0052	3	57067.11111	145600
9.877	10000000	0.0052	4	32100.25	81900
9.877	10000000	0.0052	5	20544.16	52416
9.877	10000000	0.0052	6	14266.77778	36400
9.877	10000000	0.0052	7	10481.71429	26742.85714
9.877	10000000	0.0052	8	8025.0625	20475
9.877	10000000	0.0052	9	6340.790123	16177.77778
9.877	10000000	0.0052	10	5136.04	13104

Figure.8 represents the relation between the Euler Loads and the column lengths in the same time between the Critical Loads of Buckling and the column lengths when side sway is prevented in the frame due to the loads P.

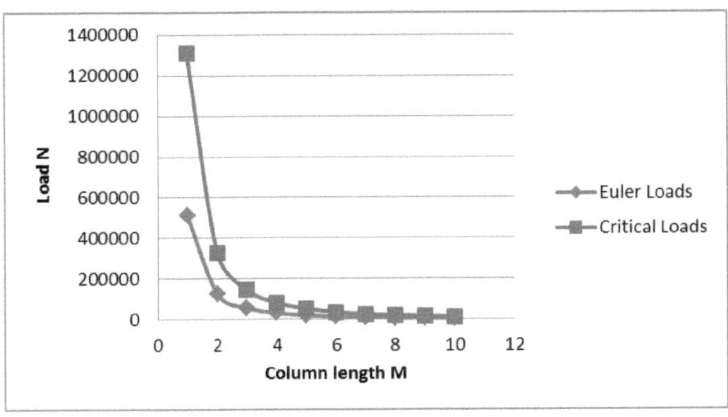

Figure.8 Euler Loads and Critical buckling loads

In Table-2, the eigenvalues of 10 mode shapes of buckling for 10 multiple column lengths of the frame are stated that are indications of the critical loads of buckling for each mode shape with respect of slenderness ratio.

Table – 2 Eigenvalues and mode shapes

	mode shape 1	mode shape 2	mode shape 3	mode shape 4	mode shape 5	mode shape 6	mode shape 7	mode shape 8	mode shape 9	mode shape 10
Length 1 M	506.69	519.46	669.23	669.82	781.09	782.8	-4166.7	-4166.7	-4166.7	-4166.7
Length 2 M	234.41	261.34	415.64	423.72	590.72	602.45	709.55	711	787.99	789.75
Length 3 M	123.24	142.58	249.17	258.38	387.97	400.13	498.32	502.65	595.07	600.67
Length 4 M	74.292	87.524	161.79	169.63	270.22	280.94	367.27	372.27	457.4	463.25
Length 5 M	49.205	58.536	111.83	118.06	196.01	204.97	279.04	283.93	361.99	367.68
Length 6 M	34.827	41.665	81.146	86.038	146.51	153.78	214.88	219.24	286.35	291.45
Length 7 M	25.889	31.081	61.297	65.178	112.95	118.86	169.35	173.13	230.58	235.07
Length 8 M	19.975	24.038	47.814	50.942	89.399	94.26	136.3	139.56	188.86	192.79
Length 9 M	15.867	19.125	38.27	40.832	72.294	76.33	111.55	114.35	156.54	159.96
Length 10 M	12.901	15.569	31.291	33.42	59.565	62.956	92.752	95.164	131.48	134.45

The eigenvalues of each mode shape associated with certain slenderness ratio of the frames with side sway prevented can be simulated in curves showing the relation between them which is in quadratic relation for the first 6 mode shapes and irregular relation for the last 4 mode shapes due to the negative values of the mode shapes for the critical loads, Figure.9 and Figure.10 represent that relation for the mode shape 1 and the mode shape 10 simultaneously.

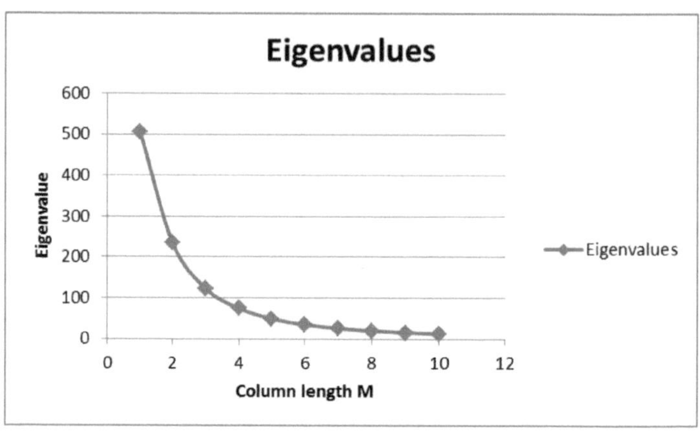

Figure.9 Eigenvalues of mode shape 1

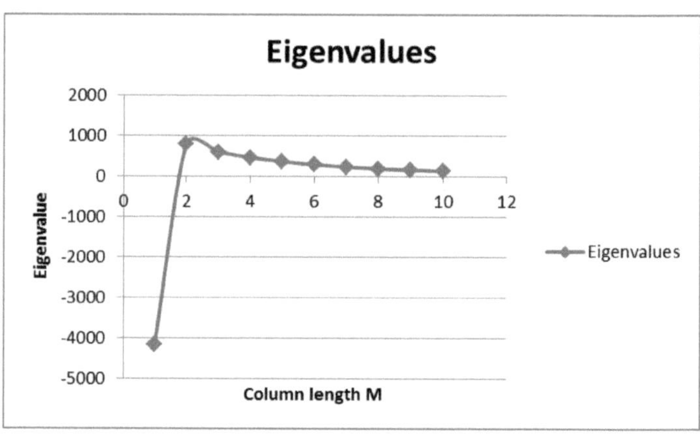

Figure.10 Eigenvalues of mode shape 10

2- Side sway buckling

If the frame is free to translate laterally at the top, it will buckle as indicated in Figure.11. The critical load for a fixed base portal frame whose beam has the same stiffness as the columns and that is laterally free to move is:

$$P_{cr} = 7.34 \frac{EI}{L^2} \qquad \dots\dots\dots\dots (6)$$

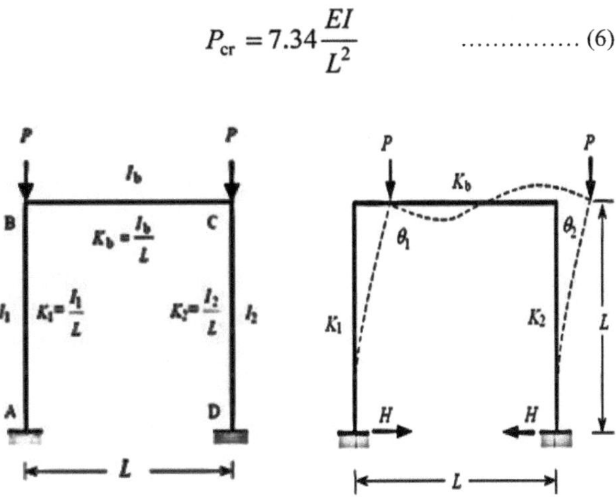

Figure.11 Sidesway buckling

The mode shapes of buckling of the frame when side sway is permitted can be seen very apparent in the output of Abaqus jobs indicating the magnitude of critical load of buckling (eigenvalues) associated with each mode shape in the Figure.12-a till Figure.12-k.

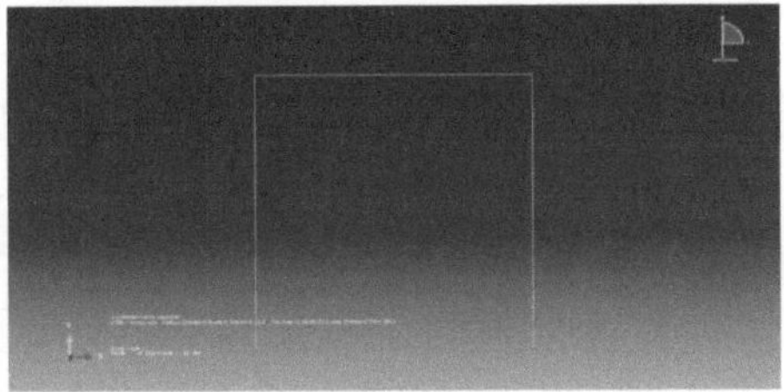

Figure.12- a un-deformed shape

Figure.12- b Mode shape 1

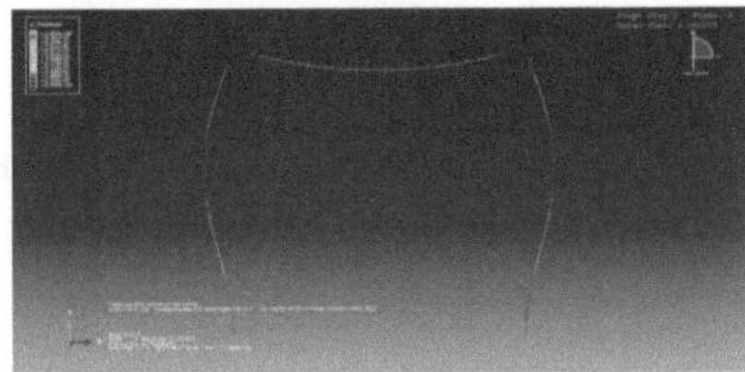

Figure.12- c Mode shape 2

Figure.12- d Mode shape 3

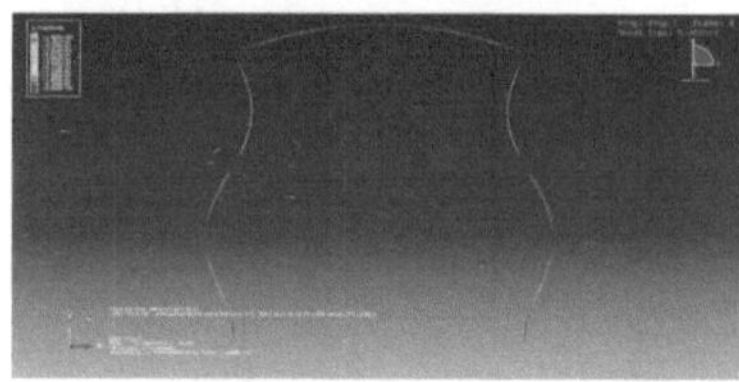

Figure.12- e Mode shape 4

Figure.12- f Mode shape 5

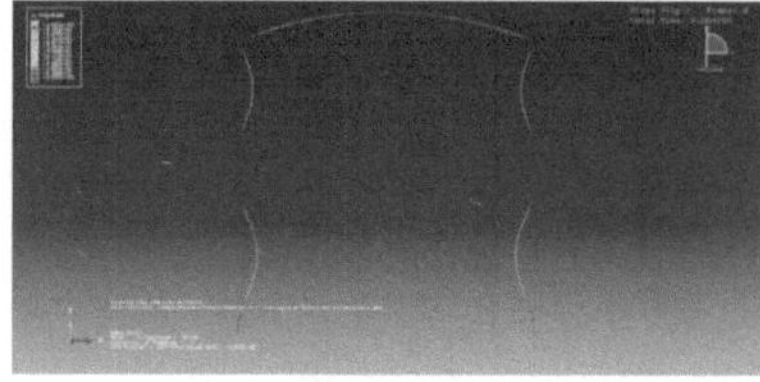

Figure.12- g Mode shape 6

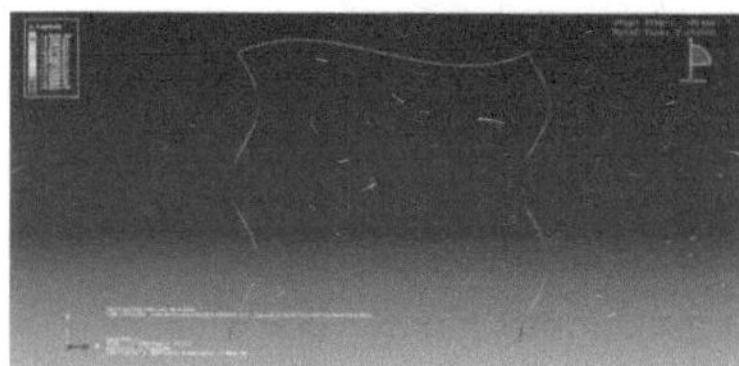

Figure.12- h Mode shape 7

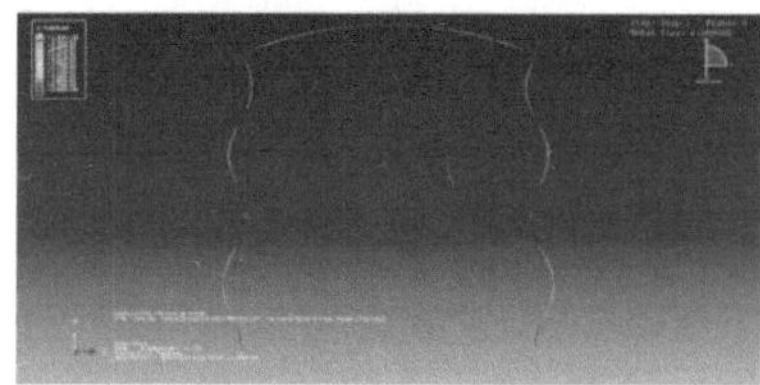

Figure.12- i Mode shape 8

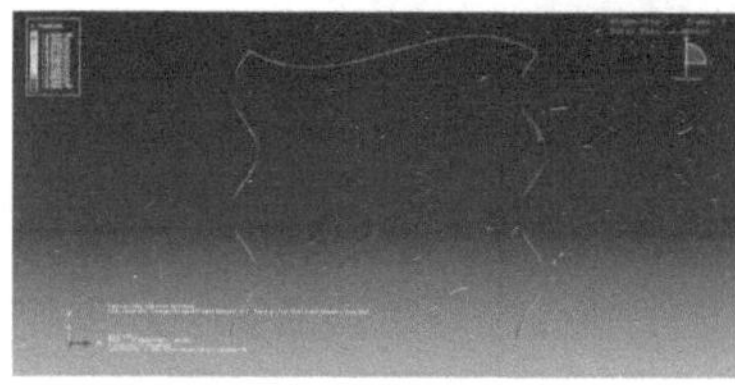

Figure.12- j Mode shape 9

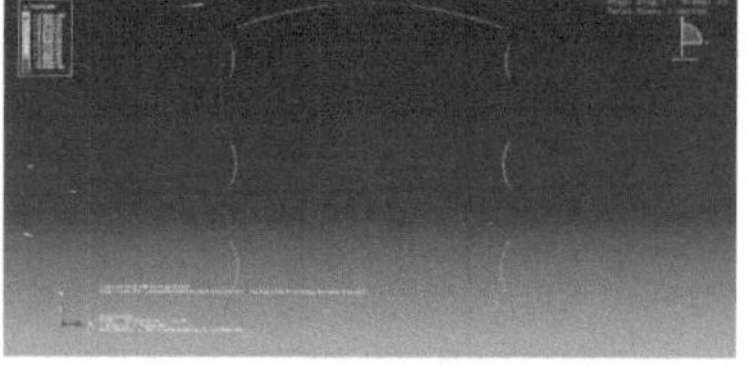

Figure.12- k Mode shape 10

Table-3 is the results of Euler loads and Critical buckling loads for 10 situations of column length for the frame with side sway permission.

Table – 3 Euler Loads and Critical Loads

π2	E	I	L	PE	Pcr
9.877	10000000	0.0052	1	513604.00	381680
9.877	10000000	0.0052	2	128401.00	95420
9.877	10000000	0.0052	3	57067.11	42408.88889
9.877	10000000	0.0052	4	32100.25	23855
9.877	10000000	0.0052	5	20544.16	15267.2
9.877	10000000	0.0052	6	14266.78	10602.22222
9.877	10000000	0.0052	7	10481.71	7789.387755
9.877	10000000	0.0052	8	8025.06	5963.75
9.877	10000000	0.0052	9	6340.79	4712.098765
9.877	10000000	0.0052	10	5136.04	3816.8

Figure.13 represents the relation between the Euler Loads and the column lengths in the same time between the Critical Loads of Buckling and the column lengths when side sway is permitted in the frame due to the loads P.

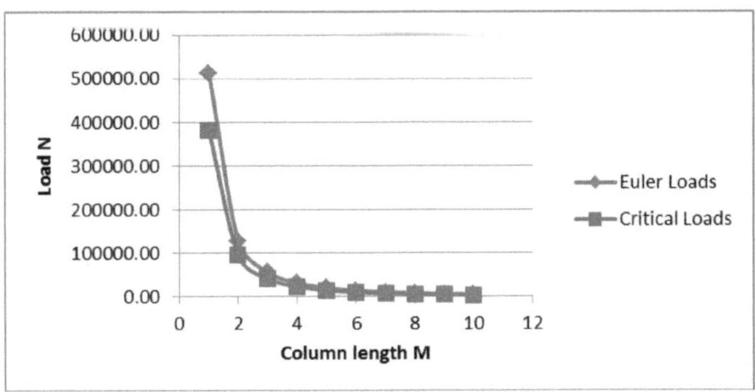

Figure.13 Euler Loads and Critical buckling loads

In Table-4, the eigenvalues of 10 mode shapes of buckling for 10 multiple column lengths of the frame are stated that are indications of the critical loads of buckling for each mode shape with respect of slenderness ratio.

Table – 4 Eigenvalues and mode shapes

	mode shape 1	mode shape 2	mode shape 3	mode shape 4	mode shape 5	mode shape 6	mode shape 7	mode shape 8	mode shape 9	mode shape 10
Length 1 M	220.44	505.53	529.13	665.52	707.5	779.95	795.66	-4166.7	-4166.7	-4166.7
Length 2 M	80.568	234.14	263.24	413.9	451.38	589.65	607.82	707.55	727.45	786.88
Length 3 M	39.3	123.16	143.32	248.58	275.11	387.62	402.22	497.56	511.96	594.55
Length 4 M	22.901	74.264	87.926	161.54	180.77	270.06	282.23	366.9	378.82	457.15
Length 5 M	14.906	49.193	58.789	111.71	125.9	195.93	205.87	278.83	288.93	361.84
Length 6 M	10.448	34.821	41.839	81.085	91.786	146.46	154.43	214.76	223.07	286.26
Length 7 M	7.7199	25.886	31.209	61.262	69.549	112.92	119.37	169.27	176.16	230.52
Length 8 M	5.9324	19.973	24.135	47.793	54.369	89.382	94.654	136.25	142	188.82
Length 9 M	4.6993	15.865	19.202	38.257	43.583	7.283	76.648	111.51	116.35	156.51
Length 10 M	3.8134	12.9	15.631	31.282	35.675	59.558	63.217	92.73	96.834	131.46

The eigenvalues of each mode shape associated with certain slenderness ratio of the frames with side sway permitted can be simulated in curves showing the relation between them which is in quadratic relation for the first 7 mode shapes and irregular relation for the last 3 mode shapes due to the negative values of the mode shapes for the critical loads, Figure.14 and Figure.15 represent that relation for the mode shape 1 and the mode shape 10 simultaneously.

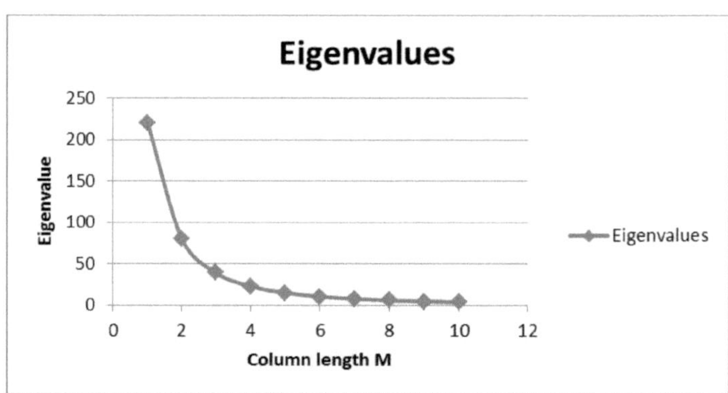

Figure.14 Eigenvalues of mode shape 1

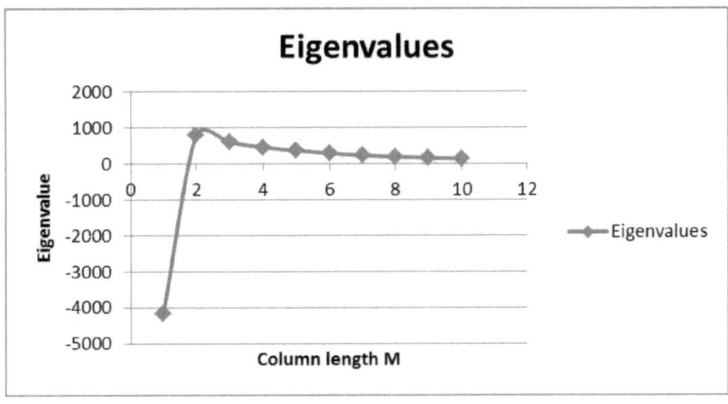

Figure.15 Eigenvalues of mode shape 10

Conclusions and recommendations

In this paper the results of the buckling analysis drawn from Abaqus jobs for the steel frames with side sway prevented and the others permitted associate with 10 multiple configuration of slenderness ratio of the columns are shown below:

1- The basic results of buckling analysis are critical coefficients (eigenvalues) and buckling mode shapes (eigenvectors) corresponding to these critical coefficients. Each critical coefficient is the factor by which the loads of appropriate load case should be multiplied to obtain appropriate loss of stability which is the buckling mode shape.

2- Negative critical coefficient (eigenvalues) for some buckling mode means that the loads of appropriate load case should have opposite direction to result in such buckling mode shape. In practice buckling mode shapes with negative critical coefficients (eigenvalues) should be neglected.

3- The theoretical values of buckling critical loads for each frame type are approximately the same value of drawn values of the critical loads obtained from buckling analysis of Abaqus jobs that are equal to multiplying the magnitude of the P applied on the frame by the eigenvalue of each buckling mode shape, and the deference between the two values could be narrowed by making mesh refinement in each seeding step in Abaqus jobs until reaching the designated theoretical value.

4- Changing the slenderness ratio of the columns to higher values by shortening the columns length every time , leads to higher values of eigenvalues for each buckling mode shape and as a result to get higher values of critical loads needed to create buckling in each type of frames especially in the steel frames that are prevented from side sway buckling.

I recommend the following points for the future works in this area which have been arisen during my research work and need more details and analysis:

1- Studying steel frames with more than one story and one bay to search the results and compare them to relevant results.
2- Mesh refinement of the models to investigate the efficiency of the results and control the output of the parameters of buckling analysis.
3- Creating steel frames models in 3D forms, and application of the same analysis steps, and drawing conclusions about the effects of this change on the analysis.

Acknowledgements

First, thank to my God(Allah) for granting me the power and faith to proceed with this research work, and special thanks to my former supervisor, Associated Professor Dr. Choong Kok Keong from the Universiti Sains Malaysia who taught me the theoretical background of this subject during my master study.

References

1- Shruti Deshpande, Buckling and Post buckling of structural   components, MSc thesis, University of Texas,chap.3, 2010.

2-Paul A. Lagace, The Column and Buckling, Massachusetts Institute of Technology, 2009.

3- The University of Sydney, Buckling of Columns, Lecture Notes of Mechanics of Solids, Chap.13, 2012.

4-Sameh Mahfouz, Design Optimization of Structural Steel work, PhD thesis, University of Bradford, chap.3,1999.

5- John E. Akin, Buckling Analysis, Rice University, chap.12, 2009.

6- B.A. Izzuddin, Simplified Buckling Analysis of Skeletal Structures, Structures & Buildings 159 Issue SB4, 2006.

7- Joseph A. Yura, The Effective Length of Columns in Un braced Frames, AISC Engineering Journal, 2003.

8- Alexander Chajes, Principles of structural stability theory, Prentice Hall, New Jersey, 1974.